ERIE COLORADO

A COAL TOWN REVISITED

BY
ANNE QUINBY DYNI

Anne Q Dyni
May 2001

© Anne Quinby Dyni 2001

ISBN# 0-9708943-0-9

All rights reserved. No part of this book may be reproduced in any form or by any means without permission in writing from the Town of Erie, PO Box 750, Erie, Colorado, 80516.

First Edition: 2001

Published by
The Town of Erie, Colorado

Cover Photo: The intersection of Briggs and Moffat streets, early 1900s. In center is the Thomas Hunter building, later the Wilson-Larson store, both of which sold groceries and general merchandise. The I.O.O.F. Lodge met upstairs. *DeWitt Brennan collection.*

CONTENTS

TOWNSITE MAP, 1871 iv

INTRODUCTION v

Chapter I
SETTLEMENT BEGINS 1

Chapter II
LAYING THE GROUNDWORK 5

Chapter III
SETTING UP SHOP 9

Chapter IV
ERIE IN THE 1890s 13

Chapter V
ALL IN A DAY'S WORK 19

Chapter VI
THE PEOPLE OF ERIE 23

Chapter VII
ENTERING THE 20th CENTURY 29

Chapter VIII
THE GOOD TIMES 37

Chapter IX
THE HARD TIMES 43

Chapter X
SPECIAL OCCASIONS 45

CONCLUSION 47

APPENDIX ONE: ERIE TIME LINE 49

APPENDIX TWO: MAJOR ERIE AREA MINES AFTER 1900 53

SOURCES 55

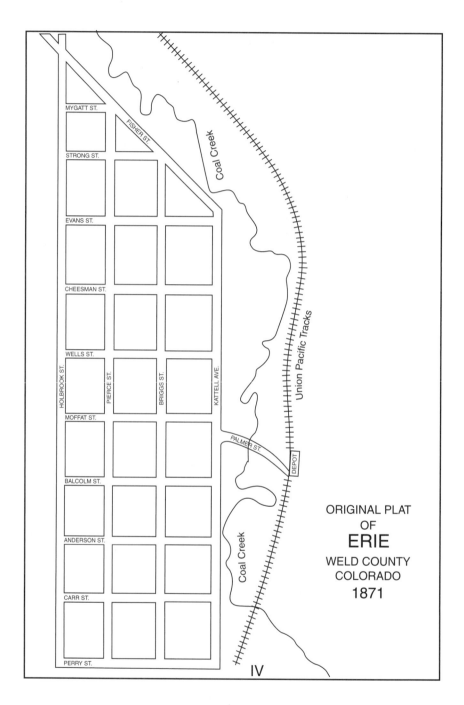

INTRODUCTION

Settlement had begun along Coal Creek when Erie was platted in 1871, and some landowners were already extracting surface coal for their own use or hauling it by wagon to nearby communities.

All that changed with the arrival of the Union Pacific Railroad Company which owned vast amounts of land in the northern Colorado coal fields. With the railroad came mining companies eager to tap into the deposits of the region. The Boulder Valley Coal Company opened the Briggs Mine in 1871, and others soon followed. Miners poured in from the eastern United States and from Europe to live and work in Coal Park, as the area was referred to on early maps.

With mining came commerce, and by 1875 many businesses had sprung up along Briggs and Kattell streets. Erie had 600 inhabitants by 1872, making it the third largest town in Weld County.

The ensuing decades were laced with labor disputes as unions clashed with management over wages and working conditions. The Knights of Labor, Western Federation of Miners, United Mine Workers and I.W.W. all entered the fray as workers' rights were threatened and company profits were jeopardized. The growing hostility boiled over on November 21, 1927, when the Colorado militia opened fire on striking miners inside the gates of the Columbine Mine. Five miners died that day and thirteen were injured. A sixth victim succumbed later in the week.

Anne Quinby Dyni
2001

CHAPTER I

SETTLEMENT BEGINS

Prior to the 1870s and the development of the northern coal fields, settlers living along Coal Creek eked out their living as farmers or merchants. The few communities existing along the Front Range were miles apart by stagecoach or horseback.

Throughout the 1860s, stagecoach companies delivered freight, mail and passengers to destinations along the foothills. The Overland Stage traveled from Denver to Laramie each day, following much the same route as the present U.S. Highway 287. Contact with distant friends and relatives improved with the coming of the railroad to Erie in 1871. Even after its arrival, however, travelers heading to Longmont had to continue on by stagecoach or spring wagon. Those bound for Valmont or Boulder detrained at the Erie terminal and boarded a stage to their final destination.

Social interaction among early settlers was limited, and families traveled great distances to attend church, school or grange meetings. Many relied on circuit-riding preachers to bring them the gospel once or twice a month. Such a man was Reverend Richard Van Valkenburg, civic leader and a founder of Erie, Colorado. Having spent many years as a Methodist preacher in the coal towns near Erie, Pennsylvania, the Reverend thought it fitting to bestow that name on its western counterpart.

The original plat for Erie was filed in 1871, following establishment of the Briggs Mine, the first commercial coal mine in Weld County. Until that time, surface coal had been extracted only on a limited basis by local landowners who delivered it to customers by horse and wagon.

It was also in 1871 that the Union Pacific Railroad extended a spur westward from Hughes Station (near Brighton) on its main line between Denver and Cheyenne. Coal from the Erie deposits was needed to fuel their huge steam locomotives. The Boulder Valley Railroad, as it was called, opened up the northern coal fields for development. Soon coal from Erie mines was being shipped by rail to markets in Denver and as far east as Kansas City.

Erie was platted by officials of the Boulder Valley Coal Company who named some of the streets after themselves; Ransam *Balcom*, Judge *Kattell* and John *Wells*. Henry *Briggs* owned the first Boulder Valley mine, and Cordelia *Briggs* was married to Reverend Van Valkenburg. Significant public figures were honored as well. Denver banker David *Moffat* had financed extension of the Union Pacific Railroad from Cheyenne to Denver, and John *Evans* was the governor of Colorado Territory.

Although pride was taken in naming Erie's streets, the founders never marked them. Some townsfolk never knew their home address until street signs were installed in the 1940s.

The main street of Erie as it appeared in 1910
Sallie McWilliams Gorce collection

Reverend Richard Van Valkenburg established the first church and Sunday school in Erie. He was a charter member of Erie's Odd Fellows and Masonic lodges.
Portrait and Biographical Record of Denver and Vicinity

Tom Richards first store where he sold farm equipment and general merchandise. *DeWitt Brennan collection*

The entire student body poses in front of Erie's first
schoolhouse which was built in 1881.
DeWitt Brennan collection

Alice Hunter Sutton poses for the camera
on Briggs Street across from the first town hall. 1910.
Lois Regnier Waneka collection

Chapter II

LAYING THE GROUNDWORK

Early photographs of Erie reveal trees and hitching posts lining a dusty Briggs Street, with wooden shops crowding the edges of its board sidewalks. On the side streets, small square houses were set back from the street, some behind picket fences. Sheds and outhouses could be seen in their back yards. A total of 100 homes had already been constructed by 1873, and building lots measuring 25 x 150 feet could be purchased for $60.

In this panorama photo, we look down Pierce Street on the
left and Wells Street to the right
DeWitt Brennan collection

At the time of incorporation in 1874, five temporary trustees governed the town until a mayor and city council were elected. Within a short time, civil matters were being conducted in a new town hall, which doubled as a public meeting place and a classroom until the schoolhouse could be built.

Keeping the peace was accomplished with the creation of strict ordinances, the appointment of a town constable and construction of a jail. An application for a post office was submitted in 1874, and for years it existed behind a counter in the rear of Thomas Richards' general store.

Tom Richards second store on Pierce Street. The post office boxes were behind his back counter. *DeWitt Brennan collection*

Although Erie's location along a stream was typical of early prairie towns, Coal Creek was not the source of its drinking water. At first, residents drew water from the Union Pacific well on the south side of town, which had been dug to provide water for their steam locomotives. Many households used mine waste water for their weekly laundry. In 1883, Erie purchased additional water from an agricultural reservoir, but with no means of treatment, its use was limited. Two years passed before a bond issue provided sufficient funding to dig three artesian wells at Holbrook and Cheesman, Moffat and Pierce, and at the south end of Briggs Street. Hydrants dispensed water at all three locations. Eventually, Erie installed a water treatment plant and the wells were abandoned in the 1940s.

For many years, Erie endured the periodic flooding of Coal Creek. Little was done to mitigate the problem, however, until the flood of 1920 prompted the town to build a dike along Kattell Street. The very next year, residents heard the fire bell warning them of a cloudburst in Coal Creek Canyon that again threatened the town. They trusted the new dike to hold. When it didn't, businesses along Briggs Street experienced flood waters up to their countertops and several homes floated off their foundations. Stella Wilson Lee described the scene years later. "It washed out railroad tracks and bridges. Days after the flood you could see muddy carpets hung over poles laid across tree branches to dry."

Pharmacist John Probert maintained his general store from 1895 to 1910 when his children took it over as Wingers Hardware Store. *DeWitt Brennan collection*

Erie's first cemetery was located on Holbrook Street. It too suffered the effects of occasional flooding and was ultimately moved to its present location on the hill east of town. The Union Pacific had donated the land in 1882 so that one of their own employees could be buried there. Today, the tombstones offer a history lesson of Erie's coal mining years. Many of the graves belong to victims of mining accidents and the influenza and diphtheria epidemics of the early 1900s.

The first schoolhouse in Erie, built in 1881, sat on the corner of Wells and Holbrook streets, the site of today's town hall. As it became overcrowded, additional classrooms were added on both sides. *DeWitt Brennan collection*

It's clear from this picture that the first schoolhouse was becoming overcrowded. *DeWitt Brennan collection*

CHAPTER III
SETTING UP SHOP

As mining companies tapped the rich coal seams of the northern fields, workers poured in from the eastern United States and from as far away as England, Wales, Scotland and Ireland. As was often the case, wives and children stayed behind until their husbands and fathers established themselves in their new jobs.

Merchants in Erie opened businesses along Kattell and Briggs streets to provide the basic services found in every frontier town... livery stable, blacksmith shop, mercantile, saloon and boarding house. But unlike most prairie towns that supplied the agricultural area around them, Erie existed primarily for the coal mines. Its general stores stocked all the tools and supplies needed by miners who were expected to provide their own picks, augers, breastplates and shovels. For $2 a day, these same miners could rent rooms at the Erie House hotel at Wells and Kattell streets. For a time, it was the only two-story building in town. Rent included meals and the use of the wash house and stables in the back.

More families arrived, swelling Erie's population to 600. Some came by horse and wagon with all their worldly possessions. Others loaded their belongings into Union Pacific railroad cars for the trip west. The influx of women and children resulted in a few changes downtown. Ella Leyner opened her millinery shop, Mr. Conway established a jewelry store, and Seidler's clothing store began selling yard goods and patent leather shoes.

By 1881, Erie's children were attending the new four-room frame schoolhouse at Holbrook and Wells streets. If increased enrollment dictated the need for additional space, classes were held in the front room of the "coffin" house, a storage facility for burial coffins on Pierce Street.

Determined to make their community thrive, many early citizens held public office in addition to holding regular jobs. The enterprising Reverend Van Valkenburg, a town trustee, also operated the Erie House hotel; Samuel Southard, the town clerk and assessor, owned a livery stable; and treasurer John T. Williams managed a general store, the only one in Erie that sold pharmaceuticals.

A stone depot stood across Coal Creek where the Union Pacific tracks skirted the east side of town. There, agent C.F. Wallace operated a telegraph service and supervised the incoming and outgoing mail.

This early blacksmith shop was probably located in Erie, although its exact location is unknown.
Charles Waneka collection.

Mike Brennan's saloon on Briggs Street, ca 1916.
DeWitt Brennan collection.

St. Scholastica was the second church to be built in Erie. *From an early postcard.*

The vacated Presbyterian Church was moved from across the street to become the Methodist Church's activity center. It later became the town community center. *Lois Regnier Waneka collection.*

CHAPTER IV

ERIE IN THE 1890s

In 1890, Erie's population still stood at about 600. Many businesses had come and gone and the town had weathered several organized strikes by the turn of the century. Most lasted less than two weeks, however, and the resiliency of Erie residents helped the town to rebound each time.

Citizens kept abreast of world and local news with the town's first weekly newspaper, the *Erie-Canfield Independent*. It was joined years later by the *Erie Independent*, but both ceased publication in 1896. The *Erie Review* attempted to fill the void, but it too quickly faded from the scene. It wasn't until 1900 when Walter McAnally began printing the *Erie Herald* that the town at last had a permanent newspaper. It continued for forty-eight years.

Erie had its share of saloons, but from the very beginning there were churches as well. The strong religious influence of its founders resulted in the establishment of a Welsh Methodist, Methodist Episcopal, Catholic, and later a Presbyterian Church. The Methodists were the first to hold services in town beginning with a Sunday school founded by Reverend Van Valkenburg.

Parishioners worshipped in the schoolhouse until each denomination could erect its own house of worship. James L. Wilson, who constructed many of the houses in town, was hired to build the Methodist Church in 1888. Years later, his daughter Stella Lee recalled how he had to begin all over again after strong winds blew the framing down. When the congregation outgrew their little church, plans were made to enlarge it. Instead of hiring another carpenter, however, the little Pleasant View church was hauled into town for an annex. Still needing room, the vacated Presbyterian church was dragged onto the property from across the street. That annex later became the Erie Community Center.

St. Scholastica was founded as a Catholic missionary church in 1898 by Father Cornelius Enders. Its first sanctuary was a recycled farm building which had been hauled into town from Canfield by Mike Brennan. As the congregation grew, it too was enlarged.

St. Scholastica Catholic Church was built in one year, due largely to the efforts of Reverend Father Cornelius and his parishioners. They gathered to dedicate the sanctuary on August 18, 1899.
Mary Young collection.

Local societies and lodges were formed throughout the 1870s and 1880s. The Good Templars was the first to organize and by 1895, the Odd Fellows, Knights of Pythias, Pythian Sisters and the Masons all had active lodges.

Erie's Knights of Pythias Lodge, organized in 1887, gave up its charter in 1943. *Erie Historical Society collection.*

The Rose of Sharon Rebekah Lodge organized the same year that Erie was incorporated. Charter members included Cordelia Van Valkenburg, Sarah Probert and Elizabeth Richards. Meetings were held in the Odd Fellows Hall. 1902. *Erie Historical Society collection.*

Gradually, the Union Pacific removed itself from the business of coal mining but retained control of its land along the railroad. Eastern financiers leased existing mines and opened new ones, leaving very few coal mines under local ownership.

The Knights of Labor moved into Colorado in 1878 and founded the state's first labor union, Local #771, in Erie. W.B. Edwards was a charter member and described the great care taken to conceal meeting locations. Members received directions to their clandestine gatherings from furtive encounters on street corners.

The Union Pacific was no longer the only railroad serving Erie. In 1881, a narrow-gauge train known affectionately as the "baby railroad" began hauling coal from the Erie-Canfield mines to Mitchell where it connected with the Denver, Utah and Pacific line into Denver. It also provided convenient passenger travel between Erie and Longmont. The line lasted for only eight years, however, before being forced out by larger railroad companies.

Later, the Chicago Burlington and Quincy line passed through Erie on its route between Denver and Lyons. At that time, a new depot was built on the south end of Pierce Street to replace the old stone building east of Kattell Street. It was later moved to High Street where it could serve both the Union Pacific and the CB&Q railroads at the point where the two tracks converged. The CB&Q continued to carry both freight and passengers until 1957, when the last "special" passenger car passed through Erie filled with Campfire Girls on an excursion to Lyons.

Erie as it appeared in 1910 from atop Cemetery Hill east of Town.
Sallie McWilliams Gorce collection.

The Union Pacific introduced motor cars in 1909, allowing passengers to
travel comfortably between Denver, Brighton, Erie and Boulder.
It was discontinued in 1925. *Dudley Pitchford collection.*

Miners from the Boulder Valley Mining Company's State Mine, 1929. *Gus Nies collection.*

This 1985 photo shows the Egnew Hotel at Briggs and Balcolm streets after it had become a residence. In its heyday it had nine sleeping rooms upstairs and a dining room below. *Erie Historical Society collection.*

CHAPTER V

ALL IN A DAY'S WORK

"When you got big enough to carry a dinner bucket without it bouncin' along the ground, you were big enough to go down in the mine," Joe "Cotton" Fletcher explained. "I went down in the mine when I was fifteen. The mine boss wouldn't give you a job. He'd give somebody permission to *take* you down. That's the way you'd get the job."

In the 1880s, Erie miners earned about $21 for a sixty-hour week, and days were ten hours long. But according to Fletcher, working conditions improved to $3.75 for an 8-hour day until the outbreak of W.W.I. After the war, wages rose once again then plummeted to $5 during the Depression. Many of the mines closed at that time. Merchants extended credit to unemployed families until conditions improved. Unfortunately, not all businesses survived this financial burden and were forced to close.

Bachelor miners lived in boarding houses like the Erie House or later, the Egnew Hotel. Rent was paid by the month for room and board, which included packing a dinner bucket to take into the mine. "They'd always give you a couple of sandwiches and somethin' sweet, and some fruit," said Joe Fletcher. "After about 40 years, it got so you'd throw your sandwich away and eat the pie, drink the coffee."

The men rose long before sunrise in the winter in order to get to work by 7:00 A.M. At the end of the shift, they walked home in the dark, the sweat of the day's work freezing on their backs. For some, a tub of hot water awaited them at home, but those living in boarding houses had to clean up in the wash house before sitting down to dinner. They were responsible for doing their own laundry.

Some mines provided housing for their non-union workers. The Columbine Mine was one of the largest in the area and employed about 350 men. Its mine camp at Serene maintained a post office, dance hall, casino and schoolhouse. Virginia Amicarella remembered winters in her family's home there when "everything froze, including our shoes which froze to the floor." The Amicarellas planted a garden and raised chickens behind their house, but were required to buy all their clothing, groceries and hardware at the company store. Sallie McWilliams Gorce described the commissary at the Puritan camp as a large general store where miners' families were obliged to shop. "They paid in scrip," she recalled. "The money for groceries was taken out of their checks so they had no opportunity to shop anyplace else."

The Columbine Mine camp at Serene provided various conveniences for its mine families, including this band stand. *Lafayette Miners Museum collection*

Almost everyone else lived and traded in Erie, Canfield or Lafayette where grocers itemized purchases on a tablet, payable at the end of the month.

John T. Williams stands in his store, surrounded by shelves of groceries and dry goods. *DeWitt Brennan collection.*

The Garfield Mine #2 south of Erie closed in 1905.
Lois Regnier Waneka collection.

John and Sarah Probert pose with their children Annie, Harriet and Willie.   *DeWitt Brennan collection.*

Scottish born Sarah Bailey and her children lived in Akron, Colorado, while her husband William worked winters in the Erie mines. The family moved to Erie in 1884.   *Margaret Russell collection.*

CHAPTER VI

THE PEOPLE OF ERIE

The heart of any community is its people, and Erie was a melting pot of diverse nationalities. Although first settled by miners from England, Scotland and Wales, the town soon saw an influx of workers from central and eastern Europe. Families living in French Town, a two-block section at the north end of Erie, found themselves isolated from the others and eventually left town for more tolerant surroundings. After 1900, many migrant families worked summers in the nearby sugar beet fields. Some put down roots and stayed to work in the mines as well.

Who were the early settlers in Erie and why did they come? Some came to minister to the ill and injured, some came to teach, and others became shopkeepers supplying the needs of the community. But most of them came to work in the coal mines.

Richard Van Valkenburg came from Pennsylvania as a Methodist Episcopal minister intent on spreading the gospel in Wyoming and Colorado Territory. Once he had helped to establish Erie, he also assumed the role of town father, civil servant, and businessman.

John Probert left Wales for the coal mines of Ohio and Illinois at age seventeen. When he heard of the opportunities opening up in Colorado, he headed west. Wishing to leave a life of mining behind him, he worked as a salesman in John T. Williams' general store. He subsequently obtained his pharmacy license, the 38th to be issued by the state of Colorado, and in partnership with Williams, opened Erie's first drug store.

Tom Richards, also of Welsh descent, had worked in mines since the age of twelve. He continued mining after arriving in Erie with just $70 in his pocket. Preferring to be his own boss, he eventually opened a mercantile supplying lumber, groceries and farm machinery to the growing community.

James Wilson came from New York State and quickly found work building houses for many of the families moving to town. He later built the Methodist Church on Holbrook Street. Wilson was instrumental in forming the all-volunteer J.L. Wilson Fire Hose Company. He would have been dismayed to learn that years after his death, a drunken member of the firehose team fell off the burning roof of the Wilson house which was then occupied by his daughter Stella Wilson Lee.

James and Charlotte "Lottie" Wilson.
Melvin Larson collection.

William Whiles was a cashier in the Erie Bank in 1911.
He worked his way up to president and remained with
the bank for 50 years.   *Mary Young collection.*

William Whiles left England in 1889 with his younger brother to live with their aunt and uncle who operated the Garfield mine. After working ten-hour days in the mine, William spent his evenings studying stationary engineering through a correspondence course. Upon graduation, he became hoist engineer at the Garfield. When the Erie Bank opened in 1903, he was hired as teller and worked his way up to president. Whiles endeared himself to many when he prevented bank robbers from escaping with $15,000 more than the $2300 they actually stole. All the money was later recovered.

Joe McWilliams moved to Weld County to manage the commissary at the Puritan Mine camp. It was there that he met his future wife Anna Zimmerman. When the couple moved to Erie, Joe opened a meat market on Briggs Street. He lost the store in the 1930s, however, when credit he extended during the Depression was never repaid. Until that time, said his daughter Sallie Gorce, "Joe McWilliams fed a lot of people in that town."

Two dollars were deducted from each miner's wages for medical care. There was always a doctor on the mine payroll who tended to their illnesses and injuries. In addition to these obligations, a mine doctor could treat private patients as well. Because of this heavy patient load, most of them relied on midwives to assist them in childbirth cases. Sarah Probert, wife of the town pharmacist, was one of the first. Dudley Pitchford's grandmother Julia Gordon was another.

"(She) was a practical nurse and midwife and attended a lot of births," he explained. "She would go to their houses on farms or in mining camps and be there when the child was born...stay for ten days possibly."

These dedicated women were sometimes paid, but more often than not it was with groceries or a grateful "thank you". Erie's earliest physician was Dr. C.P. French, the first in a long line of beloved doctors in town.

The James Wilson family.   *Melvin Larson collection.*

Erie chapter of the Ladies of the Maccabees of the World.
July 4, 1910.   *Mary Young collection.*

Erie's firehouse team often competed in races with neighboring communities.. *Erie High School collection.*

Budd Pitchford's 1915 membership certificate in the James L. Wilson Hose Company . *Dudley Pitchford collection.*

CHAPTER VII

ENTERING THE 20th CENTURY

In an attempt to entice more families to move west, the Union Pacific painted a glowing picture of life on the frontier. Their 1906 brochure described Erie as "a metropolis and nucleus of the mining industry". Those who ventured to Colorado were surely surprised to find a dusty town of 600 residents with four stores, two hotels, a school, two churches and an opera house. Still, the quality of life in Erie had improved considerably in thirty years. The town now had its own power plant and newspaper, the Kuner-Empson pickle salting station along the tracks south of town was providing jobs for many people; the two-story Lincoln School had been completed; a local bank was opening; and the telephone had come to town.

Mabel Drinkwater was a telephone operator in Erie's early years of phone service. On her night shift, she slept on a cot in the office. *Lois Regnier Waneka collection.*

Lincoln School was built in 1907, but overcrowding required that an addition be built in 1920. Since there was no auditorium, end-of-school programs and the senior play were held in the community hall.   *DeWitt Brennan collection.*

Grades 1 through 8 attended Lincoln School. Prior to building a high school, three years of secondary school were offered as well.   *DeWitt Brennan collection.*

In the 1930's, school buses like the one behind these children, were kept in a garage behind Lincoln School.
Gus Nies collection.

The Wilson-Larson general store
was owned by Fred Wilson and Jean Larson.
Melvin Larson collection.

Briggs Street, looking south.
Erie Historical Society collection.

Briggs Street, looking north.
Erie Historical Society collection.

The local government, although understaffed, still managed to keep up with its responsibilities. Their most overworked employee was undoubtedly the constable who was Erie's only maintenance man. Budd Pitchford lit the street lights each evening and maintained the dirt streets. He graded them in the summer and cleared them of ice and snow in the winter.

As the town constable in 1916, Budd Pitchford had many maintenance duties in addition to "keeping the peace".
Dudley Pitchford collection.

Roads outside of town were maintained by county road bosses and were surfaced with red ash from the mine dumps. Although the ash proved quite durable, it created considerable dust, and housewives scrambled to gather their wash off the line before four o'clock when the mines let out.

Pharmacist Charles Elzi and Dr. Charles Bixler both arrived in Erie in 1912, and left lasting impressions on the town. Elzi came directly from pharmacy college in Denver to open the only drugstore he would ever own. At the time, Erie was in the throes of a union strike that had begun two years before and would last for another two. Elzi served as mayor in the 1940s, was active in the Lions Club, and for a time was president of the Erie Bank. Folks fondly recall the delicious sodas and sundaes served from his marble soda fountain with its beautiful mirrored back bar.

These shops lining the west side of Briggs Street include Elzi's Drugstore to the left, the Erie Bank, Winslow's grocery and Dr. Bixler's office   *Dudley Pitchford collection.*

Dr. Bixler came with his wife and two daughters to open his Erie practice. Strapped for money, he had to lease a horse and cart from the local livery stable for his initial rounds. He was assured a salary from both Erie and Lafayette mining companies, however, so his standard of living improved considerably. Anna Zimmerman was Dr. Bixler's nurse during the Columbine Mine strike in 1927. She recalled the day of the shooting when wounded miners were carried into his office and lay bleeding on the floor.

A cooking class at Erie High School, 1920
Gus Nies collection.

High school students performed the operetta
"Rose of the Danube" in 1935. *Dudley Pitchford collection.*

The Puritan ball team stands beside Mike Brennan's store, one of its sponsors. *DeWitt Brennan collection.*

An earlier team strikes a more casual pose. Note the difference in uniforms. *Lois Regnier Waneka collection.*

CHAPTER VIII

THE GOOD TIMES

One of the early ordinances enacted by Erie's town council forbade any kind of racing within the city limits. This presented no problem to the Davises, however, who operated the race track south of town. People gathered from miles around to wager their hard-earned money on both horses and bicyclists.

For the kids, adventure lay just beyond the Erie town limits. There were skunks and muskrats to trap, snakes to catch and flowers to pick. North of town, a bridge over Coal Creek led to caves which extended several yards back into a sandstone outcrop. Many left initials carved in the stone as a mute record of their exploits.

In the summertime, the dirt streets became playgrounds for "Kick the can" and "Run sheep, run". Broomsticks and balls created from wadded up stockings worked just fine for impromptu baseball games. After the turn of the century, local merchants sponsored adult ball teams. Mike Brennan and Budd Pitchford sponsored the Puritan team which played down on the Coal Creek flood plain. Spectators parked their cars around the perimeter of the diamond to watch the games. Said Daisy Swallow years later, "We had some good teams. We'd holler our heads off for Erie and watch the fights after the games."

Activities in the 1880s centered around the roller skating rink south of the Odd Fellows Hall on Briggs Street. Besides the occasional concerts and recitations, folks could skate from 7:30 until 9:30 in the evening for just twenty-five cents. The floor was then cleared and couples danced to live music until midnight.

In 1885, the *Erie-Canfield Independent* announced an upcoming skating contest to determine the fastest two-mile racer in Weld County. "John Crawford, the well-known fast skater of Longmont will be there", the editorial read.

Budd Pitchford raced this bicycle competitively on the streets of Erie in his youth. *Dudley Pitchford collection.*

Dances at the turn of the century included steps that have long since been forgotten. Folks attending the Erie Leap Year Club dance in 1904 swayed to a waltz quadrille, Varsovienne and Chicago Glide in addition to the more familiar two-step, schottish and polka. "People would go down there pretty well slicked up," Joe Fletcher recalled. "Pert' near all the men that danced had a good suit".

Square dancers, too, kicked up their heels to live music at Jimmy Smith's barn, the Jarosa barn near Canfield and Mumford's barn east of Lafayette. Chaff sifting down from the hayloft added just the right amount of polish to the dance floor. Such events attracted couples from all over Boulder and Weld counties and lasted until the wee hours.

When the activities annex next to the Methodist Church was converted to a community center in the 1920s, dances were often held there. The addition of benches and chairs easily transformed it into a movie theater as well, where kids could watch Tom Mix and Tarzan movies for a dime. DeWitt Brennan sold popcorn there as a boy and recalled Pauline Phillips and Lucille Carter playing piano interludes while the film reels were changed. Years later, DeWitt married his sweetheart Billie Treasure in the same hall.

School athletic competitions and theater performances brought the entire community together. It made little difference if folks had children in school at the time. In baseball and basketball, Erie's youth could hold their own. Erie High School dominated girls basketball when its team won the State Championship in 1929. That was the same year the first high school building was completed. The boys basketball team remained undefeated in 1935 until the final playoffs.

The Erie High School girls' basketball team in front of Lincoln School, ca 1925. *Erie Historical Society collection.*

Athletic competitions continued beyond high school as Erie's volunteer firemen competed at the 1935 State Firemen's Convention in Loveland. *Dudley Pitchford collection.*

The senior play in 1914 was "Queen Esther". A small senior class that year necessitated using younger students in the cast as well. *Erie Historical Society collection.*

The first Erie High School was built in 1929. Speed ball, a game similar to soccer, was played by the boys on a dirt athletic field. Photo taken in 1944. *Lois Regnier Waneka collection.*

Erie's first airmail delivery, May 19, 1938.
DeWitt Brennan collection.

Bill Grimson's threshing crew harvests wheat on the Grimson farm south of Erie. *Gus Nies collection.*

The William Drinkwater family lived on South Briggs Street. William was a miner, but tended a garden in the summer to supplement their food supply. *Lois Regnier Waneka collection.*

CHAPTER IX
THE HARD TIMES

In a community almost totally dependent upon the coal mining industry, any ripple in the economy was immediately felt by miners and their families. Joe Fletcher described the Depression years. "It hit the miners awful hard here 'cause nobody was minin' coal, and the mines was stayed shut down."

Red Cross volunteers during W.W.I. included Mabel Drinkwater, Irene Brennan, Doris Elzi, Ruth Wilson and Evelyn Forman. *Lois Regnier Waneka collection.*

Daisy Bracegirdle Swallow had experienced labor disputes both as a young girl growing up in Erie and as the wife of coal mine foreman George Swallow. "Times were hard," she recalled. "Strikes, strikes, strikes. It was an awful struggle to get by."

Merchants carried unpaid customer accounts from month to month until work resumed and bills could be paid. Andy Deborski remembered those years. "You got so deep in the hole during the summer months without workin', it took you all winter to get out."

During the Depression, the Federal government created the Works Progress Administration (WPA) to provide jobs for the unemployed. "There was work with the WPA," according to Dudley Pitchford. "Enough to get by." WPA projects in Erie included street work and the construction of new tennis courts north of the old high school.

Even during good times, the mines ceased operations in the summer. The soft lignite coal of the northern fields was too unstable to store for long periods of time and had to be mined only during the months it was needed. In order to survive these periods of unemployment, miners earned what they could by thinning sugar beets or harvesting crops for local farmers. Their wives did housework and took in laundry while the children gleaned coal for the kitchen stove from the railroad tracks where it had fallen from passing coal cars.

Joe Fletcher felt lucky. "I had an uncle that was a section boss up here on the Union Pacific. In the summer, he'd give me a job on the section workin' on the railroad...$2.80 a day for ten hours. If the two of you couldn't put in sixteen ties and tie 'em up, why you got a cussin'."

Mike Brennan supported his family in the summertime by obtaining the bingo rights for northern Colorado. He ran bingo games at fairs in several communities north and east of Erie. His sons contributed to the family income as well with jobs at the salting station and a *Denver Post* paper route.

CHAPTER X

SPECIAL OCCASIONS

Every community along the Front Range celebrated holidays like the Fourth of July and Christmas. But because it was a mining town, Erie celebrated other occasions as well.

April 1 was "John Mitchell Day" when homage was paid to the president of the United Mine Workers who had helped to negotiate the eight-hour day for miners. School was canceled and everyone gathered for street carnivals and parades in many of the surrounding towns, including Denver. The tradition continued in Erie until the 1960s.

"Biscuit Day", an occasion no longer celebrated, was unique to Erie. Every October, baker Chris Miller gave away fresh biscuits with apple butter on the side. Folks could purchase bowls of Mulligan stew which were offered free to the ladies. Proceeds helped to support various civic projects. The ordinance forbidding races in town must have been lifted by this time since part of the day was spent betting on horse races down Briggs Street.

Labor Day was also observed as miners gathered to celebrate improved working conditions. During Prohibition Erie was considered a "wet" community, so many celebrants came from outlying "dry" areas like Longmont.

The morning of the Fourth of July always began with a parade to the ball park led by the Erie brass band, which had been around since 1875. Kids bought firecrackers, sky rockets and Roman candles at Elzi's drugstore, and the Brennan boys provided night time fireworks courtesy of the *Denver Post*. For safety sake, their father Mike ignited them in a rain trough propped against a fireplug.

The Erie concert band as it appeared in about 1920.
Erie Historical Society collection.

CONCLUSION

Closing of the Lincoln Mine in 1979 brought an end to coal mining in the northern fields. Dozens of mines had come and gone since the Briggs Mine began operating more than a century before. Erie declined as businesses left town, and a neglected downtown district did little to attract new life to the community. Residents looked elsewhere for shopping and entertainment. Only the schools, churches and lodges seemed to give Erie citizens a sense of place.

Today as Erie enters the twenty-first century, we would do well to listen to the stories of those who worked and played in this coal mining community so many years ago. The pride and determination which built Erie in the 1870s is about to be renewed.

APPENDIX ONE

ERIE TIME LINE

1871 - Boulder Valley Coal Company establishes the Briggs Mine, first commercial mine near Erie townsite.

1871 - First strike, organized at Briggs Mine to protest company rule requiring miners to screen coal before shipment. Strike unsuccessful.
- First railroad completed into northern Colorado coal fields from Hughes Station (Brighton) as a spur off of the Denver and Pacific Railroad.

1873 - "Panic of 1873" signals an economic depression forcing Erie mines to cut production because they were solely dependent on railroad consumption.
- The Boulder Valley Railroad is extended from Erie into Boulder.
- Probert building (town's first pharmacy) erected at corner of Briggs and Wells.

1874 - Town of Erie incorporated on November 16.
- Isaac Canfield attempts then abandons plans to establish a mine in the area.
- Rebekah Lodge established in Erie on October 4.
- Eureka Grange #35 established near Erie on January 19, the same year Colorado Territorial Grange established.

1875 - Erie brass band organized.

1876 - First major flood surges through Erie in May.

1878 - The Knights of Labor Local #771, first mining labor union in Colorado, chartered in Erie.

1881 - Opening run of "The Baby Railroad", along 8.2 miles of narrow gauge track between Canfield and Longmont. November 24. Survived until 1889.
- First Erie schoolhouse erected on southeast corner of Briggs and Wells streets.

1882 - Garfield Lodge #50 of the Masonic Order organized on Sept. 21.
- IOOF Lodge #46 chartered on October 18.
- Cemetery relocated to its present location on a hilltop east of Erie. Moved from original location just south of the old Presbyterian Church.

1884 - Erie's first newspaper, the *Erie-Canfield Independent*, founded. Published until 1896.

1885 - First strike by Knights of Labor. Lasted one week. Unsuccessful.

1887 - Knights of Pythias Lodge chartered in Erie on September 21.

1888 - Methodist Church dedicated.

1890 - Colorado's first United Mine Workers local is organized in Erie.
- Union Pacific ceases operation of their own mines along railroad. They lease to private concerns on a per/ton basis.

1893 - Pythian Sisters chartered.

1899 - St. Scholastica Catholic Church dedicated, August 18.

ca 1900 - Kuner-Empson and Company pickle salting station built near south end of Briggs. Closed in 1950s.
- Erie power plant built at south end of town. Began purchasing power from Public Service in 1924, and sold its distribution lines to them in 1964.

1901 - First telephones installed in a few Erie homes.

1903 - Erie Bank established. It was financially stable enough to continue in business after the stock market crash in 1929.

1907 - Lincoln School erected at Wells and Holbrook. Addition built in 1920. Deeded to town of Erie by the school district in 1966. It is now the town hall.

1910 - The United Mine Workers District #15 calls strike for higher wages in Northern Colorado coal fields. Strike continues until 1914.
- First flu epidemic hits Erie. Another to come in 1918.

1914 - Ludlow Massacre in Southern Colorado coal fields on April 20. Battle over wages kills one soldier, four strikers and several women and children.

1918 - Flu epidemic hits entire nation. Many die in Erie.

1921 - Another major flood hits Erie in June.

1922 - Strike called by UMW on April 1. Militia called out. Fourteen mines in Boulder and Weld counties involved.

1927 - Columbine Mine strike called. On November 21, militia opens fire killing 5 miners and wounding 13. A sixth miner died later in the week.

1929 - New Erie High School completed. Grades 7-12,
- High School girls' basketball team captures State Championship with a win over Las Animas, 21 to 24.

1930 - Diphtheria epidemic strikes, killing many in Erie.

1935 - High School boys' basketball team plays in the State Tournament for the first time. Loses.

1936 - Lions Club chartered on April 6.
- Erie Bank is robbed. Most of money recovered when thieves arrested at their next bank heist.

APPENDIX TWO

MAJOR ERIE AREA MINES
(After 1900)

Name	**Dates of Operation**
Boulder Valley No. 1 (State)	1917-1947
Clayton	1920-1942
Columbine	1920-1946
Eagle	1939-1978
Garfield No. 2	1886-1905
Imperial	1927-1972
Lincoln	1948-1979
Morrison	1930-1953
Puritan	1908-1939
Shamrock	1905-1956
Star	1920
Washington	1940-1967

SOURCES

BOOKS:
Smith, Phyllis. *Once a Coal Miner.* Boulder, Colorado: Pruett Publishing Company, 1989.
Erie, Yesterday and Today, Second Edition. Sociology and history classes, Erie High School. Karen Adelfang, teacher, 1974.
Portrait and Biographical Record of Denver and Vicinity. Chicago, Illinois: Chapman Publishing Company, 1898.
Colorado State Grange History, 1874-1975, Denver, Colorado: Colorado State Grange, 1975.

NEWSPAPERS AND MAGAZINES:
Erie Herald, Erie, Colorado, 5/13/1909; 5/28/1937.
Erie-Canfield Independent, Erie, Colorado, 4/17/1884; 4/24/1885; 5/29/1884; 7/18/1885; 7/31/1885; 11/27/1885; 4/9/1886; 7/16/1886; 11/19/1886; 9/1/1893; 2/8/1895.
Arner, Clifford. "Biography of William Whiles". *Erie Herald,* May 28, 1937.
Farrar, Harry. "She's Proud of Family, Proud of Erie". *Sunday Denver Post,* December 3, 1978.
Grogan, Dennis. "Unionization in Boulder and Weld Counties to 1890". *Colorado Magazine,* pp.324-341. Fall 1967.
Boulder News and Banner, Boulder, Colorado, 10/30/1885.
"The Baby Railroad", *Daily Times-Call,* Longmont, Colorado, November 23, 1981.
"Old Time Feeling: Max McAfee Recalls Erie's Heyday". *Daily Times-Call,* Longmont, Colorado, May 11, 1993.
The Erion, editorial. May 18, 1928.
Subscribers Telephone Directory, 1902. Charles Waneka collection.
Town Plat of Erie, Colorado. 1871. Erie Historical Society collection.
Weld County Cemeteries, Volume I. Lois Regnier Waneka collection.

INTERVIEWS:
Wayne Arner. 2/20/99
David Stone. 5/10/99

ORAL HISTORIES:
(Carnegie Branch Library, Boulder & Lafayette Public Library)
DeWitt Brennan, John Brennan and Dudley Pitchford
Joe "Cotton" Fletcher and Chelmar "Shine" Miller
Sallie McWilliams Gorce
Stella Wilson Lee
Elmo Lewis
Winston Morgan
Lois Waneka
Sarah Wise
Panel discussion of pioneer descendants in Erie/Lafayette.

PUBLIC DOCUMENTS:
Colorado State Business Directory, 1875. Carnegie Branch Library for Local History, Boulder, Colorado.
Colton, Roger B. and Raymond L. Lowrie. Map of Boulder-Weld County Coal Fields. 1973. USGS.
DeVischer, Lucille and Daisy Swallow. Map of Erie listing historic sites. Erie Historical Society.
Erie Business Directories, 1877, 1879, 1880. Carnegie Branch Library for Local History, Boulder, Colorado.
Subscribers Telephone Directory, 1902. Charles Waneka collection.
Town Plat of Erie, Colorado. 1871. Erie Historical Society collection.
Weld County Cemeteries, Volume I. Lois Regnier Waneka collection.

MANUSCRIPTS AND LETTERS:
Allender, Jim. "The I.W.W. Coal Strike of 1927 in Colorado". Letter to Bob Rossi. 1989. Lafayette Miners Museum collection.
Brennan, Josephine Hershey. Letter to daughter, ca 1920. Mary Young collection.
Hunt, Grace. *History of Erie.* From book codifying Erie ordinances. 1928.